Policy Framework for Investment in Agriculture

This work is published under the responsibility of the Secretary-General of the OECD. The opinions expressed and arguments employed herein do not necessarily reflect the official views of OECD member countries.

This document and any map included herein are without prejudice to the status of or sovereignty over any territory, to the delimitation of international frontiers and boundaries and to the name of any territory, city or area.

Please cite this publication as:
OECD (2014), *Policy Framework for Investment in Agriculture*, OECD Publishing.
http://dx.doi.org/10.1787/9789264212725-en

ISBN 978-92-64-21269-5 (print)
ISBN 978-92-64-21272-5 (PDF)

The statistical data for Israel are supplied by and under the responsibility of the relevant Israeli authorities. The use of such data by the OECD is without prejudice to the status of the Golan Heights, East Jerusalem and Israeli settlements in the West Bank under the terms of international law.

Photo credits: Cover © Coralie David

Corrigenda to OECD publications may be found on line at: *www.oecd.org/about/publishing/corrigenda.htm*.

Foreword

The *Policy Framework for Investment in Agriculture (PFIA)* supports host countries in evaluating and designing policies to mobilise private investment in agriculture and maximise its positive contribution to economic growth and sustainable development. Drawing on good practices from OECD and non-OECD countries, it proposes questions and guidance in ten policy areas identified as critically important for attracting agricultural investment.

Given the range and variety of relevant measures involved, the PFIA promotes policy co-ordination at host-country level. All relevant stakeholders, including not only Ministries and government bodies but also the private sector, civil society and farmer organisations, should be actively involved in a PFIA-based assessment. This tool is potentially useful for promoting agricultural investment by a wide range of stakeholders, including smallholders, small and medium enterprises, semi-commercial farmers, domestic and foreign private cooperatives and enterprises, as well as multinational and state-owned enterprises.

The PFIA draws on the *Policy Framework for Investment (PFI)* developed at the OECD in 2006 by 60 OECD and non-OECD countries. It has benefited from inputs from several OECD policy communities and has been endorsed in 2013 by the OECD Investment Committee, the OECD Committee for Agriculture and the OECD Development Assistance Committee.

OECD and non-OECD countries will continue to work together, in co-operation with the United Nations Food and Agriculture Organization (FAO) and other interested institutions and with business, labour and civil society organisations, to support the effective use and application of the PFIA.

Table of contents

Introduction

Private investment to boost agricultural production and foster food security

Private investment is essential if agriculture is to fulfil its vital function of contributing to economic development, poverty reduction and food security. Agricultural production needs to increase by at least 60% over the next 40 years to meet the rising demand for food resulting from world population growth, higher income levels and lifestyle changes. Given the limited scope for net area expansion, agricultural growth will rely mainly on new increases in productivity, supported in particular by private investment in physical, human and knowledge capital. Agricultural investment is thus critical to expand agricultural production in a context of rising land and water scarcity. It can also bolster incomes and consumption in rural areas, thereby improving global food security through enhancing not only food availability but also access to food and food utilisation.

G20, G8 and developing countries actively promote higher private investment in agriculture. In June 2012, G20 Agriculture Vice-Ministers and Deputies agreed to promote the implementation of public policies contributing to a market-oriented investment environment to encourage private sector involvement in sustainable agricultural productivity growth and farmers' market integration. In July 2009, G8 leaders called for 'increased investment in agriculture and rural development as a proven lever for combating food insecurity and as an engine for broader economic growth, prosperity and stability'. In May 2012, they committed to launch a New Alliance for Food Security and Nutrition to accelerate private capital flows to African agriculture. In addition, the Comprehensive Africa Agriculture Development Programme (CAADP), a programme of the New Partnership for Africa's Development (NEPAD) launched in 2003, aims for an annual agricultural growth rate of 6% by 2015 in Africa, notably by attracting further private investment in the sector. To achieve this objective, African countries committed themselves to allocating at least 10% of their national budgetary resources to agricultural and rural development, as stated in the Maputo declaration endorsed in 2003.

A wide range of investors are involved in the agricultural sector and policies should not only aim at increasing private investment in agriculture but also at ensuring that investments are sustainable and responsible. Domestic farmers, particularly smallholders in developing countries, are often the main source of private investment in primary agriculture. Large international investors also foster the accumulation of agricultural capital stocks in some countries. They can create employment and bring expertise, financing capacities and marketing networks to enhance the competitiveness of agricultural production and value chains. However, large-scale investments can also have adverse social and environmental impacts. Policies, laws and regulations must be well-designed and effectively implemented to ensure that such investments bring both economic and social benefits to the host country, including improved livelihoods and poverty reduction, while guaranteeing a sustainable use of natural resources.

A policy framework for sustainable private investment in agriculture

The Policy Framework for Investment in Agriculture (PFIA) aims to support countries in evaluating and designing policies to mobilise private investment in agriculture for steady economic growth and sustainable development. Attracting private investment in agriculture relies on a wide set of supply-side policies and on sector-specific public goods. A coherent policy framework is an essential component of an attractive investment environment for all investors, be they domestic or foreign, small or large. The PFIA is a flexible tool proposing questions for governments' consideration (Part I) and annotations (Part II) in ten policy areas to be considered by any government interested in creating an attractive environment for investors and in enhancing the development benefits of agricultural investment. This tool is potentially useful for promoting agricultural investment by a wide range of stakeholders, including smallholders, small and medium enterprises, semi-commercial farmers, domestic and foreign private cooperatives and enterprises, as well as multinational and state-owned enterprises. It should be noted that investors may have different needs and that enhancing small domestic investments requires specific policies.

Structure of the Policy Framework for Investment in Agriculture

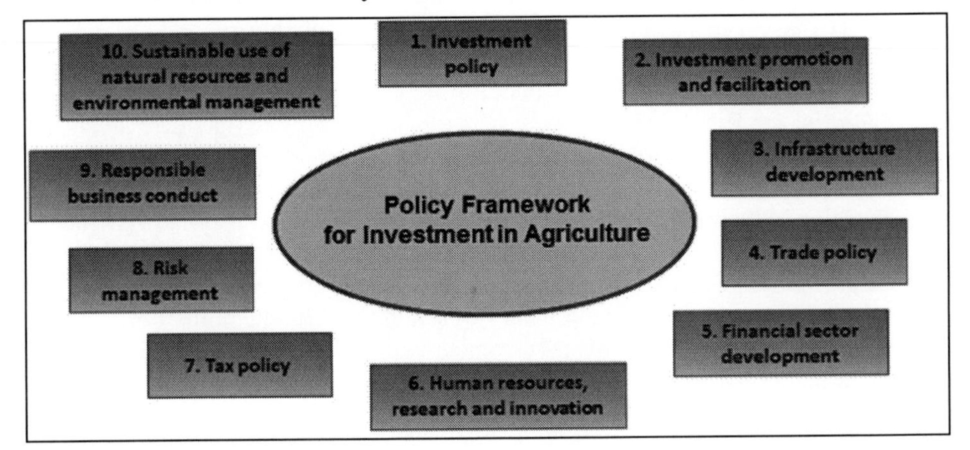

The PFIA draws on the Policy Framework for Investment (PFI) developed at the OECD in 2006 by 60 OECD and non-OECD countries. It was first developed in 2010 by the NEPAD-OECD Africa Investment Initiative, the Sahel and West Africa Club (SWAC) and the Office of the Special Adviser on Africa (OSAA) of the UN Secretary General. It has benefited from inputs from several OECD policy communities, in particular the Secretariats of the Committee for Agriculture, the Development Assistance Committee (DAC), the Committee on Fiscal Affairs and the Committee on Financial Markets. This revised version has been endorsed by the Investment Committee, the Committee for Agriculture and the DAC.

The PFIA has already been used as a self-assessment tool by Burkina Faso, Indonesia, Tanzania and Myanmar. Given the range and variety of relevant measures involved, the PFIA promotes policy co-ordination at the host-country level, both in the design and implementation phases. All relevant stakeholders, including not only Ministries and government bodies but also the private sector, civil society and farmers' organisations, should be actively involved in the PFIA process.

The PFIA can complement existing national and international initiatives aiming to attract more but also better investment in agriculture. In particular, it complements the World Bank Group's work on Doing Business by helping governments self-assess and reform a broader range of policies potentially relevant for agricultural investment. It also complements the Monitoring African Food and Agricultural Policies (MAFAP) project implemented by the FAO in collaboration with the OECD. While the MAFAP project analyses the level and composition of public expenditures in agriculture, the PFIA focuses on policy measures aimed at unlocking private investment in agriculture.

The PFIA can contribute to achieving the CAADP and Grow Africa objectives by supporting the design and implementation of regional and national agricultural investment plans and investment blueprints and by strengthening cross-sector collaboration. It can provide the Global Donor Platform on Rural Development with an instrument to facilitate donor dialogue, harmonisation and alignment around countries' priorities. The PFIA can also contribute to the New Alliance for Food Security and Nutrition launched in 2012 by the G8 as a joint initiative between African leaders, the private sector and G8 and other donors, as well as the Feed the Future initiative launched in 2009 by the US government. By promoting policy reforms for increased private investment in agriculture, the PFIA can effectively support such initiatives which aim to create an enabling policy environment for agricultural investment.

Finally, the PFIA can help implement principles for responsible investment at country-level, in particular by: promoting the FAO Voluntary Guidelines on the Responsible Governance of Tenure of Land, Fisheries and Forests in the Context of National Food Security adopted by the Committee on World Food Security (CFS) in May 2012; and by building on the ongoing consultations on responsible agricultural investment launched by the CFS in October 2011 and that should be concluded at the end of 2014. The PFIA remains a living tool which could be revised and updated based on ongoing and future work and consultations with partner countries.

Part I

Policy Framework for Investment in Agriculture

1. Investment policy

The quality of investment policies directly influences the decisions of all investors. Transparency, policy coherence and non-discrimination can boost investor confidence. Secure access to land and water and effective mechanisms for enforcing contracts and compensating expropriation are also critical to attract further investment in agriculture.

1.1. Is there an **agricultural investment strategy**? Is it aligned with food security objectives? Are sectoral policies (e.g. agriculture, education, trade, infrastructure or finance) well aligned with this strategy? Does overall investment policy support it?

1.2. What measures has the government taken to ensure that **laws, regulations and policies** for agricultural investment and their implementation and enforcement are clear, accessible, transparent and predictable and do not impose unnecessary burdens to domestic and foreign agricultural investors? Has it taken specific measures to support investors operating in the informal sector, including women?

1.3. What public consultation mechanisms, involving interested parties, in particular investors, have been established to improve **regulatory quality** in the agricultural sector, thereby enhancing the investment environment?

1.4. Are there restrictions specific to **foreign investment** in agriculture? Does the government have mechanisms to periodically review their costs against their intended public purpose?

1.5. What steps have been taken to **secure land tenure**? How are land rights allocated, administered and protected at both the central and local levels? What proportion of agricultural land has been formally registered? What measures have been taken to facilitate land rights acquisition and to provide alternatives to large-scale land transfers?

1.6. What steps have been taken to secure **access to water**? How are water rights allocated, administered and protected at both the central and local levels?

1.7. Are there initiatives to improve government capacity to **negotiate contracts** and to help officials understand the legal provisions embedded in domestic law and the country's rights and obligations under international agreements?

1.8. Is the system of **contract enforcement** effective and widely accessible to all agricultural investors? What mechanisms for **dispute settlement** have been established to ensure the widest possible scope of investor protection at reasonable cost? Are there any dispute settlement mechanisms specific to the agricultural sector, particularly as regards land tenure?

1.9. Does the government maintain a policy of timely, adequate, and effective **compensation** for expropriation consistent with international law?

2. Investment promotion and facilitation

By highlighting profitable investment opportunities and providing investment incentives, investment promotion and facilitation measures can be effective instruments to enhance agricultural investment provided they aim to correct market failures and leverage the comparative advantages of the country's agricultural potential.

2.1. What **institution** is in charge of investment promotion and facilitation? Has the government established an investment promotion agency (IPA)? Does it promote investment in agriculture and agro-processing and offer investment facilitation services at both central and local levels? Is it adequately funded and staffed to deliver its mandate and is its performance regularly monitored?

2.2. What **measures** are applied to promote and facilitate investment in agriculture, including by smallholders and informal entrepreneurs? In particular, are administrative procedures to establish a new investment streamlined and tailored to the capacity of various investors to reduce project costs? Do investment promotion and facilitation measures target specific types of investors and elements of agricultural value chains where investment is needed? Does the government undertake cost-benefit analyses to assess their impact?

2.3. Does the government intervene in **input and output markets**? Are these markets competitive?

2.4. What type of **investor-state dialogue** mechanism is in place? Does the IPA fulfil any policy advocacy role?

3. Infrastructure development

> Well-developed rural infrastructure, including good irrigation networks and transportation and storage systems as well as a reliable access to energy and information and communication technologies, can effectively attract private investors in the agricultural sector and increase agricultural competitiveness.

3.1. Are infrastructure policies **aligned** with agricultural investment objectives? How are infrastructure investment priorities identified and implemented and how are relevant stakeholders involved in decision-making?

3.2. How are **responsibilities** for infrastructure project design, provision and maintenance shared between central government and sub-national authorities?

3.3. Does the government have **clear guidelines and transparent procedures** for the disbursement of public monies for agriculture-related infrastructure?

3.4. Does the government have a clear strategy defining the roles of public and private investment in the provision of agriculture-related infrastructure? What steps have been taken to attract **private investors** to supply such infrastructure?

3.5. Does the government have a clear strategy for **irrigation infrastructure** development? How are responsibilities for the development, operations and maintenance of such infrastructure shared between government, water users and farmers?

3.6. How is the perishability of agricultural products, and thus the need for vertically integrated supply chains, considered in the development and maintenance of different modes of **transport infrastructure**, including roads, railways, ports, airports and storage facilities?

3.7. Has the government developed a strategy to ensure access to reliable and affordable **energy supply** in rural areas, including by promoting small-scale renewable energy production?

3.8. What measures has the government taken to enhance access to **information and communication technologies** by agricultural investors? Does it provide timely and accurate information on agricultural markets?

4. Trade policy

Open, transparent and predictable agricultural trade policies both domestically and across borders can improve the efficiency of resource allocation, thus facilitating scale economies, reducing transaction costs and boosting productivity and rates of return on investment. They can also help reduce price volatility and improve the stability of food markets, thereby fostering food security.

4.1. Are there any administrative, fiscal or regulatory barriers to the trade of agricultural commodities **across the country**? Do they constitute a barrier to enter agricultural markets? Is their impact quantified? Are there any measures to support the development of domestic trade?

4.2. What recent efforts has the government undertaken to facilitate **cross-border agricultural trade**, in particular by reducing regulatory and administrative border procedures? What steps has it taken to increase trade policy predictability and does it consult investors on planned trade policy changes?

4.3. Are there **specific trade measures to support agricultural investment**, such as agricultural export promotion?

4.4. Do existing tariff and non-tariff barriers to trade contribute to hindering access to **agricultural inputs and services** or raising their costs?

4.5. Are there or have there recently been **export restrictions** related to agricultural or agri-food products?

4.6. Has the government entered into bilateral or regional **trade agreements**? How effective are they in increasing market size and access? Is agriculture excluded or does it have a special treatment in these agreements? How actively is the government increasing investment opportunities through the implementation of its World Trade Organisation (WTO) commitments?

5. Financial sector development

> Efficient financial markets can allocate capital to innovative and high return investment projects of both large and small agricultural investors, thus increasing revenues and generating economic activities.

5.1. How does the **regulatory framework** contribute to a well-functioning financial market for both large and small agricultural investors? Do collateral requirements prevent some investors, including small-scale, low-capital, informal and women entrepreneurs, to access credit from formal financial institutions? If so, have any steps been taken to ease these requirements? Are there an efficient local cadastre system, a registration system for movable assets and a credit information system?

5.2. What is the **state of competition** in the formal financial sector, in particular in rural areas? What types of financial products are offered to small and large agricultural investors? Does access to credit vary by region or investor size?

5.3. How important is the role of the **informal financial sector**, including community savings, middlemen and retailers, in providing credit to farmers? How does the regulatory framework facilitate the provision of small-scale rural financial services? What is the role of microfinance and leasing?

5.4. Has the government taken any measures to facilitate **access to credit** by agricultural investors, such as by providing credit guarantees and loans on favourable terms, relaxing loan regulations or offering business development services?

5.5. Do **national and regional capital markets** play a role in raising capital for large and medium-size agricultural investors?

6. Human resources, research and innovation

> Strong human capital and dynamic agricultural innovation systems are critical to increase investment in agriculture. Policies should support high-quality education and well-functioning extension and advisory services to enhance human capital. They should promote partnerships between national and international research, better connect research with demand and effectively protect intellectual property rights to build effective innovation systems.

6.1. Has the government identified the **needs of large and small agricultural investors**, including women, and the implications of agricultural development strategies in terms of human resources and technical skills?

6.2. Do the education system and **public extension services** meet these needs, in particular by providing vocational trainings and business development services focusing on the farm as an agri-business unit rather than only on production, strengthening farmers' groups and cooperatives, and training highly qualified staff? What efforts are made to improve access to, quality and effectiveness of extension services?

6.3. How are **public research and development** (R&D) priorities defined? Are agricultural R&D institutes adequately funded and staffed? Are there measures to encourage regional R&D collaboration to share research costs and facilitate technology transfer?

6.4. Has the government taken specific measures to promote **linkages between agricultural extension and R&D** and enhance farmer-to-farmer dissemination to facilitate the rapid adoption of new techniques, practices and technologies? Does it promote local innovation and adaptive capacity? Are there specific actors and initiatives that investors can access to broaden their knowledge and skills base?

6.5. Are **investment linkages** between large agri-business companies and small and medium enterprises (SMEs) adequately promoted to foster technology and knowledge transfer? Are there mechanisms to encourage investors to train their employees and agricultural workers?

6.6. Is **private sector participation** encouraged in conducting R&D? Are there effective public-private partnerships in this area? Does the level of intellectual property protection encourage innovation by domestic and foreign agricultural investors? What is the policy regarding access to and transfer of plant genetic resources and biotechnology?

7. Tax policy

Sound tax policy enables central government and sub-national authorities to raise revenue while attracting further investment from both large and small investors.

7.1. Are the tax policy and administration in line with **agricultural investment objectives**? In particular, is the tax burden on agricultural investors appropriate to meet these objectives?

7.2. Is the tax system **neutral** in its treatment of foreign and domestic agricultural investors as well as large versus small investors? Does the government offer tax incentives to agricultural investors? Are these incentives regularly evaluated to assess their cost effectiveness?

7.3. Does the tax administration system have **sufficient capacity** to develop and implement tax policy in a transparent and efficient manner? Do tax officials support taxpayers, including small agricultural investors, in meeting tax requirements, thereby strengthening government accountability?

7.4. How is taxation administered and **co-ordinated** between the central government and sub-national authorities? Do the taxes paid by agricultural investors accrue to sub-national authorities so as to fund local public goods?

8. Risk management

As the agricultural sector faces significant weather, disease and price-related risks, effective risk management instruments can help cope with these risks, thus ensuring agricultural investors a more stable income and creating a predictable environment favourable to investment.

8.1. What policies and strategies has the government put in place to **prevent and reduce** weather, disease and price-related risks?

8.2. Do private institutions or non-governmental organisations provide **insurance** in the agricultural sector? Does the government provide support for some types of insurance? How competitive is the insurance market in the sector?

8.3. What mechanisms allow for the effective enforcement of **forward contracts**? Has the government established measures to support the development of futures markets for agricultural commodity prices?

8.4. Do agricultural extension services provide advice on **co-operative arrangements** among agricultural producers to help implement collective risk management strategies?

8.5. How does the government encourage **diversification**, including diversification in production, practices, marketing and income sources, as a risk management instrument?

9. Responsible business conduct

Policies promoting recognised principles for responsible business conduct (RBC) help attract agricultural investments that are both environmentally and socially sustainable, thereby bringing both short-term and long-term economic and development benefits to investors and host countries.

9.1. What **laws and regulations** govern RBC in agriculture, in particular as regards labour standards, tenure rights over natural resources, the right to health and anti-corruption and integrity standards?

9.2. What mechanisms are in place to **enforce** RBC laws and regulations effectively and ensure that local communities, particularly marginalised groups such as women and pastoralists, can: access timely and accurate information on proposed large agricultural investments affecting them; negotiate with large investors, in particular on access to land and water; ensure equitable benefit-sharing arrangements, formalised in a signed contract; receive fair and timely compensation if required; and settle contract disputes with investors?

9.3. Through which channels does the government **communicate** expected RBC standards to agricultural investors? How does it ensure a clear distinction between its own responsibilities and those ascribed to businesses?

9.4. How does the government **support investors' efforts** to comply with RBC laws and regulations? Does it support mutually beneficial partnerships between smallholders and large investors? Does it actively encourage private voluntary initiatives promoting social and environmental sustainability in the agricultural sector?

9.5. Does the government participate in **inter-governmental consultations** to promote recognised RBC concepts and principles?

10. Sustainable use of natural resources and environmental management

Strong and well-enforced environmental policies contribute to both promoting responsible investment and ensuring a sustainable use of natural resources such as land, soil and water, thereby fostering long-term food security, protecting biodiversity and mitigating climate change, including by providing clean energy supply.

10.1. Do existing **environmental policies, laws and regulations** effectively ensure a sustainable use of natural resources, in particular by setting clear environmental standards, requiring independent environmental impact assessments and ensuring that the pricing of natural assets reflect their true scarcity value? Do they take into account the specificities of the agricultural sector?

10.2. What **institutional mechanisms** allow for the effective implementation, monitoring and enforcement of environmental policies, laws and regulations?

10.3. Do existing policies promote access to **clean, energy-efficient and low input technologies** and encourage their adoption by large and small agricultural investors? Do they encourage investment in technologies using agricultural waste as an energy feedstock?

Part II

Annotations to the Policy Framework for Investment in Agriculture

1. Investment policy

1.1. Is there an agricultural investment strategy? Is it aligned with food security objectives? Are sectoral policies (e.g. agriculture, education, trade, infrastructure or finance) well aligned with this strategy? Does overall investment policy support it?

Promoting well-targeted agricultural investment requires a clear strategic vision that identifies key commodities and elements of their value chains – i.e. input supply, production, basic processing, trading and logistics, processing and retailing – where private investment is most needed, as well as the investors that can fill these gaps. An agriculture investment strategy should identify the major constraints to domestic and foreign investment along agricultural value chains and focus on key areas where further investment can contribute to achieving sectoral and national development objectives, including food security. For instance, investing in upstream and downstream sectors can create new markets, value-added and job opportunities and reduce post-harvest losses. An inclusive consultation process can help build consensus on the strategy and enhance the role of the private sector in agricultural development, thereby mobilising investment and reducing the potential for conflicts over specific investment projects.

Attracting private investment in agriculture relies on a wide set of policies going beyond agricultural policy. Policy coherence across various sectoral policies is thus critical not only to design more efficient policies but also to create an attractive environment for all agricultural investors. Sectoral policies influencing the business climate in agriculture include in particular trade and investment, education, research and development, infrastructure, financial market, environment and tax policies.

Policy coherence calls for political commitment and good co-ordination across the various government bodies responsible for policy design and implementation and for the participation of relevant stakeholders at the national and sub-national levels. Policy co-ordination mechanisms range from informal mechanisms to the systematic screening of legislative and policy proposals. Aid policies and strategies should also be well aligned with recipient country policies to ensure coherence.

1.2. What measures has the government taken to ensure that laws, regulations and policies for agricultural investment and their implementation and enforcement are clear, accessible, transparent and predictable and do not impose unnecessary burdens to domestic and foreign agricultural investors? Has it taken specific measures to support investors operating in the informal sector, including women?

Clear and accessible laws, regulations and policies contribute to creating a safe and reliable environment for agricultural investors. It helps businesses to assess investment opportunities on a more informed and timely basis and reduce transaction costs, thus fostering investment. Transparent information on how governments implement and change regulations as well as predictable regulations increase investor confidence. They support investment in particular by small agricultural investors facing particular challenges to entering the formal economy but also by large foreign investors who have to function with different regulatory and administrative systems.

The clarity and transparency of the regulatory framework can be enhanced in particular by: consulting with domestic and foreign investors, including farmers' organisations; strengthening co-ordination across various bodies and levels of government; informing and involving all relevant ministerial departments when preparing or changing regulations; simplifying regulations and administrative processes; developing registers of existing and proposed regulations; and publishing and disseminating regulations and policies.

As shown by the World Bank Doing Business reports, heavier regulation and excessive red tape increase costs and delays for investors and result in higher corruption levels among public officials. Regulatory Impact Analysis (RIA) examining the likely benefits and costs of regulations, including their social and environmental impacts, can help pinpoint administrative burdens and inform regulatory decision-making. Such analysis should rely on strong institutional capacity in government agencies at national and sub-national levels.

1.3. What public consultation mechanisms, involving interested parties, in particular investors, have been established to improve regulatory quality in the agricultural sector, thereby enhancing the investment environment?

Laws and regulations should be developed in an open and transparent fashion, with appropriate legislative controls and procedures for effective and timely inputs from interested national and foreign parties. The regulatory framework should address the need for promoting business growth, investor confidence and agricultural competitiveness on the one hand, and requirements of social and environmental sustainability on the other hand.

Designing a sound regulatory framework requires consultations with a wide range of stakeholders and media freedom to scrutinise these processes. Stakeholders should include in particular: relevant government bodies at all levels; potential domestic and foreign agricultural investors; agricultural workers; professional farmers' organisations; civil society organisations; communities affected by investment decisions, including marginalised groups such as smallholders and women; and wider interest groups. Consultations with umbrella organisations, in particular through innovative fora to ensure they participate in decision-making processes, can improve transparency and inclusiveness within investment processes. Effective consultations help to ensure that affected parties understand the content and the potential impact of new and proposed regulations, thus facilitating their implementation. In contrast, inadequate consultations may result in inefficient regulation and uncertainty for investors, thereby undermining the investment climate.

1.4. Are there restrictions specific to foreign investment in agriculture? Does the government have mechanisms to periodically review their costs against their intended public purpose?

The national treatment principle provides that foreign and domestic investors are treated equally and that foreign investors are not discriminated against on the basis of their origin. This implies, for instance, that non-resident investors be allowed to establish a subsidiary or branch or take participation in an existing domestic enterprise on conditions equivalent to those offered to resident investors. The national treatment principle also involves not discriminating between foreign and domestic investors regarding their operations.

Policies favouring domestic firms over foreign ones involve a cost. They can result in less competition and efficiency losses, thus hindering investment. For instance, measures restricting transfers of investment-related capital, including repatriating earnings and liquidated capital, can adversely affect investor confidence and international investment inflows. Similarly, complex land acquisition processes and insecure land rights for foreign companies can reduce their investments in agriculture. However, it should be noted that land acquisition by large foreign companies increases social risks. Partnerships with local land rights holders, such as contract farming arrangements, should be preferred over land acquisition as efficient mechanisms to enhance the sustainability of investments. In any case, the costs and benefits of exceptions to non-discrimination should be regularly evaluated to determine whether their intended purpose (e.g. protection based on the infant industry argument) remains valid.

1.5. What steps have been taken to secure land tenure? How are land rights allocated, administered and protected at both the central and local levels? What proportion of agricultural land has been formally registered? What measures have been taken to facilitate land rights acquisition and to provide alternatives to large-scale land transfers?

Secure and well-defined land rights encourage new agricultural investment and the upkeep of existing investments. Such tenure rights do not need to be ownership rights but can also be lease rights, such as tenant farming. They carry an intrinsic economic value by entitling investors to participate in any profits derived from their investment. For instance, secure land rights incentivise land owners to promote investments enhancing land productivity. Investors need to be confident that their rights are properly recognised and protected and that they guarantee against forced evictions.

Reliable land rights registration can effectively enhance tenure security by recording individual and collective tenure rights and allowing investors to seek legal redress in cases of violation of property rights. Land property registrars can also allow land rights holders to use land as collateral to access credit. As mentioned in the *Voluntary Guidelines on responsible governance of tenure of land, fisheries and forests in the context of national food security (VGGT)* endorsed by the Committee on World Food Security (CFS) in May 2012 (see section on responsible business conduct), they should be properly maintained and publicised, in particular by developing an integrated framework including existing recording systems and other spatial information systems. By raising the land value, land registration can incentivise smallholders to sell their land to outsiders, which may enable them to exploit opportunities outside the sector. However, care needs to be taken so that it does not lead to them being deprived of their livelihoods. Land registration should thus be carefully implemented to mitigate such risks.

Land rights registration should provide tenure security not only to large investors but also to small investors, such as smallholders, women and pastoralists, to have positive distributive impacts, in particular by taking into account customary rights. Policies and laws on tenure rights should be non-discriminatory and gender sensitive and should consider the fact that land has not only economic but also social, cultural, spiritual, environmental and political value to communities. Land rights registration should thus identify all existing tenure rights and rights holders through an inclusive consultation process and develop socio-culturally appropriate ways of recording customary rights of local communities (VGGT, 2012). Similarly, regulated spatial planning, which can support sustainable investment and balanced territorial development and help reconcile the different objectives of land use, should be developed through wide public participation and gender-sensitive policies and laws to ensure that priorities and interests of local communities are reflected (VGGT, 2012).

To enhance the efficiency of the land administration, the responsibilities of the central government versus local authorities should be clearly defined to promote transparency and enhance law implementation and enforcement. If accompanied by appropriate capacity-building and financing mechanisms at the local level, the decentralisation of land rights allocation and administration to local authorities can ensure higher accountability in land management and facilitate the involvement of local communities in the decision-making process and in consultations with investors, thereby enhancing transparency in land allocation decisions. The legislation can provide for ex-ante and ex-post environmental and social impact assessments (ESIAs) to ensure that land allocation follows a transparent and inclusive process across the whole territory. Certificates can also be delivered to large investors that cope with such assessments.

Land rights acquisition is often a complex and slow process for large investors and measures to facilitate land acquisition can effectively facilitate agricultural investment. At the same time, appropriate safeguards should protect legitimate tenure rights from risks arising from large-scale transactions in tenure rights. ESIAs could be made compulsory for land acquisitions exceeding a certain area. They should be conducted in a transparent manner by independent agencies and fair compensation should be paid to former land rights holders. Ceilings on permissible land transactions could be introduced and transfers exceeding a certain scale strictly regulated, particularly through open and transparent consultations with local communities. Promoting a range of production and investment models, such as partnerships with local tenure rights holders, can also offer good alternatives to large-scale land transfers of tenure rights (VGGT, 2012).

1.6. What steps have been taken to secure access to water? How are water rights allocated, administered and protected at both the central and local levels?

Similarly to land rights, secure and well-defined water rights encourage new agricultural investment and the upkeep of existing investments. Water management policies should promote a sustainable and equitable use of water resources, in particular by considering not only the economic value of water but also its social and environmental value. To be recognised as legitimate, water rights allocation should take due account of customary arrangements for access to and allocation of water and of the dichotomy between formal and informal socio-economic institutions. Mechanisms should be in place to regularly review water rights allocation. While water rights registration can secure access to water, it remains quite challenging in contexts where water users are small and geographically scattered. If poorly implemented, administration-based water rights can undermine customary arrangements and increase conflicts over such rights.

While levying water fees and levies can improve cost recovery of water management services and serve as an incentive to conserve water, assessing the real value of the water resources is particularly difficult. Furthermore, collecting fees can be expensive if water users are scattered and water withdrawals difficult to measure due to the lack of maps and infrastructure. The benefits versus the challenges of levying water fees should thus be closely analysed before implementation.

If accompanied by appropriate capacity-building and financing mechanisms at the local level, the decentralisation of water management can enhance the participation of local communities in allocating water rights and levying water fees, in particular through Water Users' Associations. It can also facilitate their involvement in the planning, construction, operation and maintenance of community-based water supply schemes. The responsibilities of the central government versus local authorities should then be clearly defined to effectively enhance law implementation and enforcement.

1.7. Are there initiatives to improve government capacity to negotiate contracts and to help officials understand the legal provisions embedded in domestic law and the country's rights and obligations under international agreements?

The international legal framework, through international investment agreements for example, provides protection to foreign investors to be considered by host country authorities when negotiating contracts. In case of a dispute, contract enforcement could be treated as treaty rights. If an investment treaty contains an umbrella clause covering contractual obligations, the host state would have an international obligation to abide by these obligations. Their violation could substantiate a claim brought by the investor against the state before an international arbitration tribunal.[1] These aspects need to be evaluated carefully by governments, especially in land-leasing contract negotiations. Capacity-building at all government levels can ensure the effective understanding and enforcement of negotiated contracts and treaties, and help clarify how domestic law interacts with international investment treaties (such as bilateral investment treaties or investment chapters of regional trade agreements) and WTO laws, thereby promoting transparency and certainty. Meanwhile, investors should comply with the legislation and participate in good faith in any contracts they sign, such as by disavowing bribery and other corrupt practices.

1.8. Is the system of contract enforcement effective and widely accessible to all agricultural investors? What mechanisms for dispute settlement have been established to ensure the widest possible scope of investor protection at reasonable cost? Are there any dispute settlement mechanisms specific to the agricultural sector, particularly as regards land tenure?

Contract enforcement is critical to gain investors' trust because, ultimately, it is the possibility of buying and selling assets through market transactions that reveals an asset value. The legal framework must ensure contract enforcement, property and tenure rights' protection and fair dispute resolution – all hallmarks of an efficient market. Confidence in market integrity can also be strengthened by developing effective, affordable and impartial dispute settlement mechanisms. As mentioned in the VGGT, these mechanisms should be accessible to all, women and men, in terms of location, language and procedures, and take active measures to prevent disputes from arising and escalating into violent conflicts.

Such mechanisms can consist of judicial and administrative bodies but also of alternative dispute settlement procedures, such as arbitration, mediation and conciliation hearings by industry bodies or specialised agencies. The latter allow settling disagreements between transacting parties at reasonable cost while recourse to the judiciary system can often be slow and expensive. Where customary or other established forms of dispute settlement exist, they should provide for fair, reliable and non-discriminatory ways of promptly resolving disputes.

1.9. Does the government maintain a policy of timely, adequate, and effective compensation for expropriation consistent with international law?

A natural corollary of the protection of property and tenure rights is the need for compensation when a government expropriates property. The challenge for governments lies in keeping their power to expropriate investors for public interest while protecting investors' interests. Despite the widespread acceptance of the need for timely, adequate and effective compensation in case of expropriation, in particular in bilateral investment treaties and recent regional agreements, the power of government to expropriate has a negative impact on the investment climate. To mitigate such negative effects, governments are encouraged to consider whether similar results can be achieved through other public policy means. If a government decides to expropriate land or other agricultural property, this decision ought to: serve a public purpose; observe due process of law; be non-discriminatory; and follow transparent rules defining not only the situations in which expropriations are justified but also the process of expropriation (identification of affected people, information and consultation at all stages, transparency) and the process by which compensation should be determined.

As mentioned in the VGGT, appropriate systems should be used for the fair and timely valuation of rights. They should promote sustainable development objectives and take into account non-market values, such as social, cultural, and environmental values. Sale prices and other relevant information should be recorded, analysed and made accessible to provide a basis for accurate, reliable and transparent assessments of values (VGGT, 2012).

2. Investment promotion and facilitation

2.1. What institution is in charge of investment promotion and facilitation? Has the government established an investment promotion agency (IPA)? Does it promote investment in agriculture and agro-processing and offer investment facilitation services at both central and local levels? Is it adequately funded and staffed to deliver its mandate and is its performance regularly monitored?

Centralising investment promotion and facilitation activities, such as information dissemination and policy advocacy, within a single agency can be cost effective and provide an opportunity to present a coherent impression of a country's attractiveness to investors. IPAs can act as a liaison between different government bodies, concerned institutions and interested investors, and provide guidance to investors while evaluating investment projects based on transparent criteria. Governments can consult well-documented best practices of IPA activities to take effective measures and avoid costly mistakes. An IPA can focus on specific priority areas and sectors, such as agriculture. Tunisia went even further by establishing an IPA focusing only on agriculture, the Tunisian Agricultural Investment Promotion Agency.[2]

2.2. What measures are applied to promote and facilitate investment in agriculture, including by smallholders and informal entrepreneurs? In particular, are administrative procedures to establish a new investment streamlined and tailored to the capacity of various investors to reduce project costs? Do investment promotion and facilitation measures target specific types of investors and elements of agricultural value chains where investment is needed? Does the government undertake cost-benefit analyses to assess their impact?

As a country establishes a sound investment environment, investment promotion and facilitation measures can be useful instruments to attract new investors, especially in small and remote markets or in countries with a recent history of macroeconomic and political instability. Effective investment promotion also serves to raise the positive image of the economy by highlighting profitable investment opportunities.

Investment promotion and facilitation measures can consist of specific incentives to encourage investors to target the agricultural sector, such as tax holidays or simplified licensing procedures through a one-stop shop. They can target specific geographic areas, types of investors, agricultural commodities, or elements of the value chain such as input production, agro-processing or marketing. They can also target specific investments that are likely to bring long-term economic and social benefits to the host country and local communities. IPAs can then play an active role in promoting business linkages between these investments and smaller domestic investors. One-stop advisory services provide investors with an easy access to information on the necessary steps to start or expand their business and speed up the granting of necessary permits and licenses, thereby saving time and reducing costs for investors. Decentralised services can facilitate and increase access to such services by all types of investors.

While these measures may serve as a partial rectification for market imperfections that cannot be addressed by direct policy reforms, they should not substitute for policy measures aiming to create a sound investment environment. Indeed, the effectiveness of tax incentives in attracting investment has not been conclusively demonstrated. Tax breaks and holidays rarely translate into increased investment if not accompanied by other reforms to improve the business climate.[3] They can also reduce the government's ability to mobilise domestic resources for development purposes, while their implementation raises administrative costs. Governments should thus carry out regular impact assessments to evaluate their actual social and economic costs and benefits. They might conclude that well-run infrastructure, strong intellectual property rights and business-friendly regulations are more effective in attracting investment.

2.3. Does the government intervene in input and output markets? Are these markets competitive?

While price support, price stabilisation, and input subsidies are used to promote agricultural investment and fulfil short-term objectives with respect to incomes and food security, they may not always address the causes of market failures and can sometimes become a drain on public finances and crowd out spending on public goods that are essential to attract private investment.

Price support for food products aim at increasing farmers' income and thus investment capacities, but it may adversely impact the majority of farm households who are net food buyers. Price stabilisation can provide a more stable investment environment and limit the impact of adverse shocks on producers, but it thwarts the development of private risk management and can export instability onto world markets. Over the long term, these market interventions treat the symptoms of market failure and under-development rather than the causes. Such objectives may be implemented through stockholding and distribution measures by state trading enterprises relying on anti-competitive practices, leading to inefficiencies and crowding out private investors. The demonstrated willingness of competition authorities to prevent, correct and sanction anti-competitive practices can have a significant positive bearing on the investment climate. The benefits and costs of price support and stabilisation should thus be judged relative to the benefits and costs of tackling the underlying problems directly through social policies in the short term and investment in public goods in the long term. Expenditures on these interventions should not crowd out essential investments in building more productive and resilient rural economies.[4]

Input subsidies provide an operationally simple and politically attractive way of addressing multiple economic objectives, including: stimulating production, offsetting high transport and input supply costs, making inputs affordable to farmers without credit, and allowing farmers to learn about the benefits of new inputs. In the short term, they can redress failings such as under-developed infrastructure, missing markets for credit and inputs, and a lack of knowledge of the benefits of using improved seed and fertiliser. However, they are an indirect, and relatively inefficient, way of addressing such objectives. They impede the development of private input markets, may have negative distributional effects and distort resource allocation, and are prone to capture by vested interests. In particular, state-owned enterprises with monopolistic market positions may capture benefits from input subsidies, thereby discouraging existing as well as prospective investors. If the objective is to address market failures, an exit strategy should be implemented once that task is accomplished.[5]

2.4. What type of investor-state dialogue mechanism is in place? Does the IPA fulfil any policy advocacy role?

Receiving regular feedback from investors and providing quick and accurate responses to their queries enables governments to develop a sound and business-friendly investment climate. IPAs can play an important role in facilitating effective communication between investors and the government. They are often the main source of feedback from investors to policy-makers and, conversely, they can be an effective communication channel for investors on government activities impacting on the business climate. Such interactions can take many forms. For instance, an IPA can act as a useful facilitator by matching foreign investors with local entrepreneurs, hosting a database of business opportunities, and advocating policies. These functions require in-house technical and managerial capacity, including a qualified staff with relevant business experience in agriculture.

3. Infrastructure development

3.1. Are infrastructure policies aligned with agricultural investment objectives? How are infrastructure investment priorities identified and implemented and how are relevant stakeholders involved in decision-making?

Adequate agriculture-related infrastructure, in particular in rural areas, can connect investors to their customers and suppliers and enable them to specialise and take advantage of new technologies, thereby increasing productivity and improving incomes. Such infrastructure should not only support primary agricultural production but also the development of agro-processing industries to reduce post-harvest losses and increase value-addition. Policy-makers should prioritise the various needs for agriculture-related infrastructure development, operations and maintenance (O&M), based on the various public policy objectives to be achieved, and allocate necessary public funding to support it. Before deciding on investing in agriculture-related infrastructure, the government should undertake a cost-benefit analysis, factoring in risks and the potential project impact. The participation of relevant stakeholders at the national and sub-national levels in the identification of priority infrastructure projects can help ensure a balance of competing interests. Investors should have access to clear and efficient communication channels to indicate their infrastructure needs to policy-makers. Co-ordination mechanisms should also be established across the various government agencies responsible for infrastructure and agriculture throughout the policy design and implementation stages.

3.2. How are responsibilities for infrastructure project design, provision and maintenance shared between central government and sub-national authorities?

Sound infrastructure project management and effective infrastructure O&M require good co-ordination and a clear allocation of responsibilities between central government and sub-national authorities. Central government can be responsible for raising the necessary funds for infrastructure projects and for policy planning and upstream project preparation, including the tendering process, while sub-national authorities can oversee project O&M. Budgetary allocations from the central government to sub-national authorities accompanied by capacity-building activities are often necessary to ensure adequate service delivery by sub-national authorities.

3.3. Does the government have clear guidelines and transparent procedures for the disbursement of public monies for agriculture-related infrastructure?

Governments should ensure that public funds are disbursed in a timely manner to meet agriculture-related infrastructure needs. Infrastructure development must be combined with adequate incentives for O&M activities, in particular at the level of sub-national authorities. For instance, the cost of O&M can be shared between the central government and sub-national authorities according to their fiscal capacity through matching grants where the size of the central government's assistance for infrastructure rehabilitation is determined by O&M funding from sub-national authorities. As maintenance activities represent a substantial share of infrastructure investment needs, especially in developing country contexts, separate maintenance funds can be established for the good planning and financing of maintenance.

3.4. Does the government have a clear strategy defining the roles of public and private investment in the provision of agriculture-related infrastructure? What steps have been taken to attract private investors to supply such infrastructure?

To meet demand and promote broad-based agricultural growth, agriculture-related infrastructure should be supplied by both the government and the private sector whose respective roles need to be clearly defined. Mobilising private investment to finance agriculture-related infrastructure development and maintenance, such as through public-private partnerships, can help ensure that public policies address the needs of the private sector and help address the dilemma of public resource scarcity. For example, the government can encourage private agricultural investors to provide rural infrastructure when negotiating land allocations with them.

Private sector participation in infrastructure development relies on: a clear and sound regulatory framework; efficient and transparent management of public funds; clear allocation of responsibilities between private investors and government; competitive and transparent contracting processes; adherence to established timetables by the public and private parties; procedural fairness to ensure a level-playing field for all investors; government support for high-impact economic and social projects that are not commercially viable; government contingent support or guarantees for certain risks; protection of investor rights; and effective dispute settlement mechanisms, such as those provided by domestic courts, international arbitration panels or independent regulatory agencies.

3.5. Does the government have a clear strategy for irrigation infrastructure development? How are responsibilities for the development, operations and maintenance of such infrastructure shared between government, water users and farmers?

While most agricultural production worldwide is rain-fed, irrigation is an efficient channel to increase agricultural productivity and reduce the risks related to chronic water shortages due to droughts or geographical constraints. The management of irrigation networks has been gradually decentralised in many countries to ensure that beneficiaries, in particular water users and farmers, participate fully in the planning, construction, O&M and management of community-based irrigation schemes. Responsibilities between various stakeholders should be clearly defined to ensure coherent and reliable irrigation infrastructure provision and maintenance. In terms of funding, shifting the financing burden of irrigation infrastructure development from government to water users can increase financial sustainability, improve governance by increasing accountability, and support the creation of public-private partnerships in irrigation infrastructure. For instance, some water user associations are already responsible for collecting water user charges to fund O&M costs at the local level.

3.6. How is the perishability of agricultural products, and thus the need for vertically integrated supply chains, considered in the development and maintenance of different modes of transport infrastructure, including roads, railways, ports, airports and storage facilities?

By enabling agricultural investors to buy and sell in different and distant markets, good transportation and logistics systems facilitate the creation of vertically integrated agricultural value chains. Fresh agricultural produce in particular require reliable and efficient transport infrastructure to link agricultural production with post-harvest handling, storage, processing and marketing, and to reduce delays and associated product losses.

Secondary roads facilitate access to processing plants and markets in urban areas by agricultural investors located in rural areas. These roads should be connected to rail and road networks and feed into cross-border corridors, including seaports and airports, to facilitate exports, allowing small-holder farmers to move into commercial farming and boost their incomes. Smallholders often use rudimentary storage facilities, leading to significant post-harvest losses. Adequate storage systems from the farm level up to national and international markets include granaries, silos, refrigeration units, on-site warehouses and strategic grain reserves. Storage can not only limit food losses and waste but also allow producers to sell at times when prices are highest and increase food security in lean times.

3.7. Has the government developed a strategy to ensure access to reliable and affordable energy supply in rural areas, including by promoting small-scale renewable energy production?

Affordable and reliable energy supply is a necessary condition for promoting the development of agricultural value chains and investment in food processing and distribution. For instance, preserving fresh agricultural products to supply national and international markets requires cold storage systems that depend on a reliable electricity supply. While back-up generator facilities can provide a more reliable supply of power than electricity from the grid in many developing countries, they are costly and only relatively large agricultural investors can afford them. The development of the grid is thus critical to ensure regular and affordable energy supply. Electricity can be generated from fossil fuels, such as oil and coal, but also from clean energy sources, such as solar, wind, hydro and geothermal power or biomass. Renewable energy is particularly promising as access to such energy can be expanded through affordable decentralised projects. Solar powered drip-irrigation can maximise the impact of irrigation while making use of a reliable and clean source of energy.

Access to energy – including renewable energy - in rural areas can be improved by supporting decentralised approaches and encouraging private participation in energy generation and distribution, especially in mini-grids or stand-alone power plants. Private participation can be enhanced by: designing a tariff structure that ensures cost recovery; promoting predictable feed-in-tariffs; providing targeted input subsidies to consumers; designing a sound regulatory framework for private participation; unbundling the energy sector by separating energy generation, distribution and transmission functions; opening the energy-generating industry to independent power producers to increase competition among several energy providers; and shifting from a fully vertically integrated monopoly to that of a single-buyer model whereby independent power producers contract with a national utility.[6]

3.8. What measures has the government taken to enhance access to information and communication technologies by agricultural investors? Does it provide timely and accurate information on agricultural markets?

Access to information and communication technologies (ICTs) contributes to strengthening agricultural value chains by enabling investors to communicate rapidly and cheaply with distant suppliers and customers. In particular, mobile phones support marketing by allowing smallholders to: access regular and reliable market information, notably on prices; conduct negotiations with suppliers and buyers without travelling long distances; make cashless transactions through mobile banking; and strengthen their bargaining positions vis-à-vis buyers. Furthermore, ICTs underpin many service sectors, such as finance, insurance and transportation, which are all critical to attract investors.

As ICTs evolve rapidly, governments need to regularly evaluate the relevance and impact of their regulatory arrangements on ICT services and prices. With the advent of wireless ICT technologies, new investments are needed to bring the infrastructure and services to under-served areas. To expand access to mobile phone technology, governments can work with private operators to provide handsets to farmers in rural areas, subsidise air time, or revise regulations to ensure that phone operators charge fair tariffs. A predictable and independent industry regulator and competition among service providers can encourage private participation in ICT services. Policies should not only aim to expand access to ICTs but also promote capacity-building to ensure that smallholders are able to use ICTs as tools to identify marketing opportunities.

4. Trade policy

4.1. Are there any administrative, fiscal or regulatory barriers to the trade of agricultural commodities across the country? Do they constitute a barrier to enter agricultural markets? Is their impact quantified? Are there any measures to support the development of domestic trade?

Smooth and reliable domestic agricultural trade can foster investment by facilitating market access and increasing rates of return on investment. Domestic trade is heavily dependent on the quality of transportation systems (see infrastructure development section) but also on policies regulating or taxing such trade.

Regulations on domestic trade can limit competition among buyers of agricultural commodities, thus reducing market and value chain efficiency. For instance, regulations can limit the number of licenses delivered to such buyers or force individual producers to sell through cooperatives, thereby reducing the number of potential buyers. Competition can also be limited by increasing entry costs for new buyers, such as by imposing them to operate both as agricultural traders and processors. Furthermore, charges and taxes imposed on inter-regional agricultural trade can represent a significant fiscal burden for traders. Roadblocks used to control the transport of agricultural products between various regions and to collect such charges and taxes cause delays resulting in heavy losses due to the perishability of most agricultural products. Regulations and taxes limiting domestic agricultural trade should thus be avoided.

4.2. What recent efforts has the government undertaken to facilitate cross-border agricultural trade, in particular by reducing regulatory and administrative border procedures? What steps has it taken to increase trade policy predictability and does it consult investors on planned trade policy changes?

Slow and costly custom procedures increase business costs, thereby lowering rates of return and discouraging investors, including smallholders. Border delays are particularly crucial in the agricultural sector as most products are perishable. Such delays can be reduced by simplifying border procedures and increasing regulatory co-operation between exporting and importing countries to harmonise requirements. The impact of customs regulatory and administrative procedures on the investment environment should be periodically assessed.

Uncertainty on trade policy may be created by unpredictable government decisions obstructing the smooth functioning of international supply chains and causing less stable pricing structures. Investors compensate for uncertainty by adjusting upwards the rate of return required to undertake an investment project and reducing their overall investment. Trade policy-making can be made more transparent via better communication of policy decisions and more regular public consultations with businesses and other constituents.

4.3. Are there specific trade measures to support agricultural investment, such as agricultural export promotion?

While home countries can promote trade through export credit schemes, host governments also have a number of policy options available. For instance, export financing and technical assistance for local suppliers to meet international standards can increase the attractiveness of the host country as an investment destination and strengthen the trading capacity of local farmers.

4.4. Do existing tariff and non-tariff barriers to trade contribute to hindering access to agricultural inputs and services or raising their costs?

Agricultural investors must have access to services, machinery and agricultural inputs, in particular seeds and fertilizers, at reasonable cost which requires not only a competitive domestic input market but also a sound trade policy. In the short term, temporary and well-targeted trade barriers may facilitate the emergence of domestic input industries and services by limiting competition. However, trade barriers - high import duties and non-tariff barriers, such as import licensing requirements, standards, labelling and local content requirements - can also hinder access to intermediate goods and raise their cost. While the WTO obligations from the Agreement on Trade-Related Investment Measures (TRIMs) prohibit measures obliging local sourcing, some countries may require a minimum proportion of investment project inputs to be sourced locally, regardless of their cost competitiveness. Policy-makers should evaluate the potential dampening effects of such measures on investment. Internationally traded services are particularly important inputs for agro-processing industries and a strategic component of global agricultural value chains. Service exports do not only benefit the host country which imports them but also make the home countries' multinational enterprises globally competitive. Liberalising trade of agricultural services may thus benefit both home and host countries.

Furthermore, food quality and safety standards, testing and certifications can limit market access to importing countries. The capacity of investors to meet these regulatory standards, including private voluntary and international standards, should be enhanced, including by strengthening metrology and standards institutions as well as accreditation, testing and certification bodies.

4.5. Are there or have there recently been export restrictions related to agricultural or agri-food products?

Open trade policies can facilitate the integration into global supply chains, thus attracting large foreign investors in agriculture. Export restrictions and taxes can send wrong signals to investors and may negatively impact investment. In particular, export bans can prohibit access to larger and often closer regional markets and thus reduce incentives to invest. They can undermine the credibility of the exporting country, discourage exporters and reduce investment in the long run. While these bans often aim to ensure sufficient domestic supply, they can have short and long-term negative effects. In the short term, agricultural investors can lose an opportunity to sell their output in export markets that may offer higher prices while, in the long term, agricultural producers may shift to less profitable crops. Export bans are often imposed in an unpredictable manner as emergency measures without consulting investors, thereby increasing uncertainty for investors. Similarly, export taxes usually aim to support domestic agro-processing or encourage supply on the domestic market at prices below world prices. Although they can be used to invest in public goods to the benefit of investors, they may also hinder investment and their costs and benefits should thus be regularly assessed.

4.6. Has the government entered into bilateral or regional trade agreements? How effective are they in increasing market size and access? Is agriculture excluded or does it have a special treatment in these agreements? How actively is the government increasing investment opportunities through the implementation of its World Trade Organisation (WTO) commitments?

By harmonising trade policies and regulations (including sanitary and phytosanitary measures), establishing free trade areas and creating larger markets, WTO-consistent bilateral and regional trade agreements can increase trade, raise the potential to exploit scale economies and allow investors to harness the benefits of the comparative advantages of the host country. They usually include investment provisions that complement domestic efforts to create a sound investment environment. By enabling enterprises to source suppliers and sell their output at globally competitive prices, they create additional investment opportunities. Such agreements can also help prevent and manage food crises by facilitating the transfer of food commodities from surplus to deficit countries.

5. Financial sector development

5.1. How does the regulatory framework contribute to a well-functioning financial market for both large and small agricultural investors? Do collateral requirements prevent some investors, including small-scale, low-capital, informal and women entrepreneurs, to access credit from formal financial institutions? If so, have any steps been taken to ease these requirements? Are there an efficient local cadastre system, a registration system for movable assets and a credit information system?

Large, vibrant and competitive financial markets with adequate prudential safeguards lower transaction costs, mobilise savings and allocate financing efficiently to investors, enabling them to seize promising investment opportunities and to better manage risks. They impose discipline on firms to perform, thereby enhancing economic efficiency. They also minimise risks of systemic instability, thus increasing macroeconomic and investment performance. The key factors supporting the good functioning of financial markets include: macroeconomic stability; competition in financial services, including from foreign enterprises; prudential oversight of risk taking; transparency; good information flow; and clearly-defined and well-enforced rights of borrowers, creditors and shareholders. When creditor rights are weak, financial intermediaries will be less willing to extend credit to firms and when shareholder rights are weak, potential shareholders will be less willing to extend equity finance.

Well-defined property rights that investors can pledge as collateral can increase access to finance and allow firms to borrow on a longer-term basis at a lower cost. Transparency on the type and value of collateral and legal security to seize it in case of default can encourage lending. The legal framework should support the use of all kinds of assets and rights as collateral, in particular by establishing publicly open, affordable and efficient registry systems. An accurate, comprehensive and accessible cadastre and land registration system can enable land rights holders to use land as collateral. While immovable assets – land and buildings – are the most common form of collateral, low-capital movable assets could also be accepted as collateral to enable borrowers without immovable assets to access finance. Establishing a registry for movable assets can provide the necessary transparency and legal security for such collateral. For instance, warehouse receipt systems can be effective ways to use movable assets as collateral. They can also reduce market risks for both traders and producers and help stabilise agricultural markets.

A credit information system can facilitate credit expansion by enabling creditors to access reliable and transparent information on borrowers, thereby reducing information asymmetry, credit risk, transaction costs, reliance on collateral and over-indebtedness. Specialised agencies such as credit bureaus can often collect such information at a lower cost than financial institutions themselves. Data protection and credit reporting laws can allow collecting creditor information and building a transparent credit information system while protecting borrower rights. Such system should not only register credit information but also the abilities of the debtors, such as their entrepreneurial ability, to select investments with the highest economic returns able to drive innovation and agricultural growth and better assess the likelihood of loan repayment.

5.2. What is the state of competition in the formal financial sector, in particular in rural areas? What types of financial products are offered to small and large agricultural investors? Does access to credit vary by region or investor size?

Banks' efficiency depends, among other factors, on the status of competition in the banking sector. A highly concentrated sector might result in lack of competitive pressure to attract savings and to channel them efficiently to investors and underserved segments of the economy or the population, such as smallholders located in remote areas. Developing countries with a limited number of large commercial banks can suffer from relatively high interest rate spreads and a lack of customised and diversified financing schemes and products. As opportunities for local financing are scarce and access to capital abroad is limited, access to finance for both large and small agricultural investors, particularly women, can thus be a challenge. Furthermore, commercial banks tend to have a stronger presence in urban areas than in rural areas, thereby limiting financing opportunities for rural SMEs that are vital to agricultural growth. Enhancing competition in the banking sector may require removing barriers to private credit expansion for agricultural investment. While accessing credit can allow firms to invest, mobilising savings is also critical to enable them to accumulate capital. Banks should be encouraged to offer savings products and payment systems in rural areas.

5.3. How important is the role of the informal financial sector, including community savings, middlemen and retailers, in providing credit to farmers? How does the regulatory framework facilitate the provision of small-scale rural financial services? What is the role of microfinance and leasing?

The informal financial sector can effectively complement the formal banking sector by providing financial services in underserved areas. The presence of non-bank players allows for greater innovation and coverage. Regulation should therefore not favour the development of one type of institution over the other.

Microfinance institutions (MFIs) in particular can provide credit to smallholders unable to access credit through the banking sector because they do not meet the requirements or live in remote areas. Regulatory standards should not be too stringent to allow for their development. Policies should aim at ensuring that they have sufficient capabilities to limit systemic risk. In addition to supporting mechanisation, leasing can ease access to finance for smallholders since: it does not require any collateral; leasing contracts can be agreed to without considering credit histories; and lessors take little risk as asset ownership remains with them. Governments may foster leasing markets not only by designing and implementing a sound legal framework but also by building capacities in the leasing industry.

5.4. Has the government taken any measures to facilitate access to credit by agricultural investors, such as by providing credit guarantees and loans on favourable terms, relaxing loan regulations or offering business development services?

Small-scale farmers are often unable to provide required collateral and are perceived as risky clients by financial institutions in particular due to their exposure to climate uncertainties influencing credit repayment. Credit guarantees can thus effectively ease their access to credit. Credit guarantee schemes mitigate risks by extending the bilateral relationship between lender and borrower to a third party, the guarantor. They usually alleviate information asymmetry as, by accepting to endorse a bank loan application, the guarantor signals confidence, presumably based on better knowledge of the project and the risk of a proposed transaction. By means of additional vetting, the guarantor helps redirect funds to underserved market segments, in particular SMEs and export-oriented businesses. Banks therefore might extend loans that they would not have granted otherwise.

The government can establish its own credit guarantee schemes but also support other actors – such as large investors with their outgrowers or Mutual Guarantee Associations (MGAs) – to become guarantors. Large investors relying on outgrowers as suppliers can act as credit guarantors to ensure outgrowers can afford high-quality agricultural inputs. Members of a MGA, usually entrepreneurs in a given industry or regional cluster, collectively underwrite a loan extended to one of them, thereby spreading the risk among all members, using the ability of borrower's peers to assess the risk better than a bank given their proximity, and reinforcing borrower discipline through peer pressure, since default has consequences for all peers.

Access to finance can be enhanced not only through supply-side measures but also by increasing demand for financial services. Low demand can be related to the fact that entrepreneurs are reluctant to use their assets as collateral which means they dilute their ownership or cede control to equity investors. But the inability of small entrepreneurs to evaluate available funding options and to understand bankers' concerns and needs also constitutes a serious barrier to demand for financial services which can be addressed through continuous knowledge and skills upgrading, including by providing financial literacy programmes. Such trainings complemented by legislation on consumer protection can also help reduce over-indebtedness.

5.5. Do national and regional capital markets play a role in raising capital for large and medium-size agricultural investors?

Capital markets enable providers and users of funds to interact without bank mediation. Transparent and liquid capital markets act as an important financing channel for both the corporate sector and governments by allowing institutional and retail investors to steer their funds to the use they deem most appropriate. In the agricultural sector, capital markets can offer obvious advantages such as: (i) mechanisms for listing and raising capital that may restructure and modernise agricultural enterprises and contribute to creating competitive value chains; (ii) trading products, such as futures and other derivatives; and (iii) information tools providing commodity-specific information enabling investors to manage their price risk (see section on risk management). In Africa, while stock markets are for the most part not sufficiently developed to cater to financial needs, automation of trading systems, regional integration efforts – as evidenced by the existing West African Regional Stock Exchange and the future East Africa Stock Exchange – and increased primary market activity could boost the size and liquidity of capital markets in the near future.

6. Human resources, research and innovation

6.1. Has the government identified the needs of large and small agricultural investors, including women, and the implications of agricultural development strategies in terms of human resources and technical skills?

Human resource development (HRD) policies should rely on a careful assessment of the skills' needs of all types of investors from those that can be addressed by technical trainings to those requiring efforts on tertiary education, focusing not only on agricultural production but also on other elements of the value chain, particularly agro-processing. For instance, identifying the needs of large investors in terms of local sourcing can allow specialised providers of business development services to provide local producers with the necessary skills to become suppliers of such investors. HRD policies should also support the implementation of agricultural development objectives by focusing, for instance, on specific agricultural sub-sectors or technical skills, and taking due account of the specific role of women in agricultural production and processing.

6.2. Do the education system and public extension services meet these needs, in particular by providing vocational trainings and business development services focusing on the farm as an agri-business unit rather than only on production, strengthening farmers' groups and cooperatives, and training highly qualified staff? What efforts are made to improve the access to, quality and effectiveness of extension services?

HRD in agriculture relies not only on primary, secondary and tertiary education but also on agricultural training institutes, public extension services, vocational trainings and business development services. While extension under decentralised conditions can be undermined by budget cuts linked to structural adjustment, decentralisation can allow extension to be more responsive to the needs of local farmers, and especially small-scale farmers. Well-trained and well-equipped extension workers considering smallholders as agri-business units are critical to: provide appropriate technical advice and disseminate new technologies; help smallholders minimise market risks and better respond to the needs of large agricultural investors, for instance by providing trainings on standards required to enter international markets; and effectively link them to input and output markets, thus increasing their competitiveness.

6.3. How are public research and development (R&D) priorities defined? Are agricultural R&D institutes adequately funded and staffed? Are there measures to encourage regional R&D collaboration to share research costs and facilitate technology transfer?

Agricultural R&D is critical not only to disseminate improved agricultural inputs allowing to increase agricultural productivity levels and enter international markets but also to develop new techniques, practices and technologies addressing the new challenges faced by agriculture, such as climate change by developing drought resistant cultivars. The Tea Research Institute in Sri Lanka is one example of a well-staffed institute that has successfully established its brand in the global market.[7] Agricultural R&D requires stable budgetary allocations to fund long-term research programmes over several years and highly qualified staff able to develop innovative technologies. Farmers should be actively involved in defining R&D priorities by identifying research needs and existing constraints to innovation and to the adoption of new techniques. They can ensure that R&D incorporates and builds on local knowledge.

Regional collaboration can effectively reduce research costs and regional initiatives have recently received further attention. For instance, the First Agricultural Productivity Programme for Eastern Africa 2009-2015 aims at enhancing regional specialisation in agricultural research, increasing collaboration in agriculture training, and facilitating the transfer of agricultural technology, information, and knowledge across national boundaries. Harmonising and strengthening seed laws and regulations across countries and eliminating barriers to regional seed trade can incentivise increased regional collaboration on R&D by enhancing the dissemination of seed research. It can also allow investors to benefit from the best seed technology at competitive prices.

6.4. Has the government taken specific measures to promote linkages between agricultural extension and R&D and enhance farmer-to-farmer dissemination to facilitate the rapid adoption of new techniques, practices and technologies? Does it promote local innovation and adaptive capacity? Are there specific actors and initiatives that investors can access to broaden their knowledge and skills base?

R&D agencies should collaborate effectively with agricultural training institutes and extension services to adapt new techniques, practices and technologies to local contexts and ensure a wide and rapid dissemination of these technologies to agricultural investors and workers. As most of them are location specific, their adoption requires decentralised and participatory approaches, not only involving research institutes, universities and the private sector, but also farmers and local non-governmental organisations. Local institutes can thus be established to adapt these techniques, practices and technologies to local needs and disseminate agricultural knowledge and information. Simultaneously, technology users, including agricultural investors, should also be able to influence the choice of research programmes to ensure that they respond to existing demand and to improve research monitoring. Interlinked national and local innovation systems can stimulate and support local initiatives and provide them with the opportunity to scale-up and replicate innovative technologies.

6.5. Are investment linkages between large agri-business companies and small and medium enterprises (SMEs) adequately promoted to foster technology and knowledge transfer? Are there mechanisms to encourage investors to train their employees and agricultural workers?

Large investors can bring new technology and knowledge into the local economy. However, certain measures should be in place to incentivise these investors to train local agricultural producers and agri-businesses and ensure they acquire the necessary technology to meet their requirements and become their suppliers, in particular by complying with sanitary and phytosanitary standards (SPS) and specific production practices.

Government policies can encourage investors to endow their suppliers with new skills and technology through co-financing arrangements, tax incentives and subsidies. SME promotion measures for skills upgrading and access to finance can also help create investment linkages by increasing the absorptive capacity of SMEs, thus ensuring they benefit from technology and knowledge transfer from large investors. An IPA can be well-placed to promote these linkages by helping investors find suitable partners and transmitting HRD needs to relevant public authorities and institutions. Governments can also provide guidance to investors to build partnerships with their suppliers, such as outgrower schemes, resulting in more competitive local suppliers. In addition, global reputational reputational risk provides large investors with an incentive to elaborate supplier development programmes to remain competitive on developed countries' markets.

6.6. Is private sector participation encouraged in conducting R&D? Are there effective public-private partnerships in this area? Does the level of intellectual property protection encourage innovation by domestic and foreign agricultural investors? What is the policy regarding access to and transfer of plant genetic resources and biotechnology?

In a context of scarce financial resources, the active participation of the private sector in agricultural R&D should be encouraged not only to increase R&D funding and opportunities but also to ensure that research programmes better meet private sector demand. Private companies could be allowed to produce, distribute and market agricultural seeds. There are several successful examples of private participation in R&D. In Australia, for instance, the Rural R&D Development Corporation (RDC) is a public-private partnership between the government and the industry. They both set its priorities and fund its operations with the government collecting compulsory levies from the industry and providing matching contributions. In Tanzania, the private sector is heavily involved in R&D on export crops through the Tea Research Institute of Tanzania (TRIT), the Tobacco Research Institute of Tanzania (TORITA) and the Tanzania Coffee Research Institute (TACRI). Governments can draw from best practices to design successful public-private partnerships that are effective in developing and disseminating innovation.

Well-defined intellectual property rights (IPRs) and enforcement mechanisms can encourage innovation and its diffusion by the private sector. Strong protection does not only favour local R&D by foreign investors but also provide SMEs with a secure environment to invest in innovation, and thus contribute to local entrepreneurship and competitiveness. In particular, compliance with the Agreement on Trade-Related Aspects of Intellectual Property Rights (TRIPS), adopted by the members of the WTO in April 1994, could be improved. Article 27.3(b) of the Agreement underlines that patents should be available 'for any inventions, whether products or processes, in all fields of technology, provided that they are new, involve an inventive step and are capable of industrial application'. WTO members should provide for the protection of plant varieties either by patents or by a system created specifically for the purpose ("sui generis"), or a combination of the two. However, they may 'exclude from patentability plants and animals other than micro-organisms, and essentially biological processes for the production of plants or animals other than non-biological and microbiological processes'. While strengthening IPRs, governments should ensure that IPRs do not hinder access to valuable technologies by small investors who cannot afford expensive technologies.

Governments should also facilitate access to and transfer of plant genetic resources. Indeed, contracting parties to the International Treaty on Plant Genetic Resources for Food and Agriculture agreed to 'establish an efficient, effective and transparent multilateral system in order to facilitate access to plant genetic resources for food and agriculture, and to share, in a fair and equitable way, the benefits arising from the utilisation of these resources, on a complementary and mutually reinforcing basis'. Similarly, one of the objectives of the Convention on Biological Diversity is the fair and equitable sharing of the benefits arising out of the utilisation of genetic resources, including by appropriate access to genetic resources and by appropriate transfer of relevant technologies.

7. Tax policy

7.1. Are the tax policy and administration in line with agricultural investment objectives? In particular, is the tax burden on agricultural investors appropriate to meet these objectives?

Taxation provides a predictable and stable flow of domestic revenue to governments allowing them to finance development objectives and to provide public goods, such as infrastructure. Governments face the challenge of striking the optimal balance between creating a business and investment-friendly tax regime and leveraging sufficient revenue for public service delivery which, in turn, makes economies more attractive to investors.

Agricultural investment is influenced by the level of taxation. Absolute and comparative assessments on competing tax jurisdictions influence location decisions of foreign investors. Taxes in the agricultural sector include in particular import duties on agricultural inputs and export taxes on agricultural commodities (covered in the trade section), income tax, value-added tax and country-specific taxes, such as produce cess in Tanzania paid by buyers of farm products to local authorities as a percentage of farmgate prices. In deciding the tax burden, governments should consider the objectives of agricultural investment and overall tax policy, including efficiency and equity, compliance costs and revenue requirements. Most smallholders do not pay income taxes as they fall under the threshold for raising such tax. Thus, a relatively high income tax rate may incentivise them to remain small to avoid paying such tax, thereby discouraging investment and growth.

7.2. Is the tax system neutral in its treatment of foreign and domestic agricultural investors as well as large versus small investors? Does the government offer tax incentives to agricultural investors? Are these incentives regularly evaluated to assess their cost effectiveness?

Tax systems may impose a non-uniform effective tax rate on businesses, based on criteria such as the size, ownership structure (e.g. domestic versus foreign), business activity or location of the enterprise. Policy-makers should examine the arguments in favour and against such differential tax treatment to be able to justify it as costs of tax differentiation might be higher than benefits. Where justifications are weak, consideration should be given to a non-discriminatory approach.

Tax incentives can encourage investors to target the agricultural sector or to invest in specific agricultural sub-sectors or regions. They have to be used carefully (see section on investment promotion and facilitation). In fact, investors are generally willing to accept a higher tax burden in exchange for a more attractive business climate offering a sound and transparent regulatory framework, skilled labour, secure access to inputs or adequate infrastructure. In order to effectively attract investment, governments should thus focus more on certainty and consistency of tax treatment, the avoidance of double taxation and efficient tax administration than on tax incentives.

7.3. Does the tax administration system have sufficient capacity to develop and implement tax policy in a transparent and efficient manner? Do tax officials support taxpayers, including small agricultural investors, in meeting tax requirements, thereby strengthening government accountability?

While tax policy should be in line with agricultural investment objectives, tax administration should be transparent and efficient to ensure effective policy implementation and facilitate access to tax information by all investors to inform their investment decisions. Transparent tax policy implementation can also enhance the credibility and accountability of the public sector vis-à-vis its citizens and the business community.

This requires in particular good capacity and co-ordination between various government institutions involved in tax administration and collection at the national and sub-national levels. Bribery and other corrupt tax practices should be fought as they increase transaction costs and distort competitive conditions, thus hindering investment.[8] Inefficient tax administration can also discriminate against small investors. For instance, tax exemptions on agricultural exports can be poorly enforced for small investors that are not registered and cannot claim tax reimbursement. Efficient tax collection should rely not only on good governance standards but also on supporting taxpayers by simplifying tax payments and instructing businesses, in particular small businesses, on how to handle tax filing and documentation.

7.4. How is taxation administered and co-ordinated between the central government and sub-national authorities? Do taxes paid by agricultural investors accrue to sub-national authorities so as to fund local public goods?

Policy co-ordination between central government and sub-national authorities is crucial for efficient and transparent taxation. Local tax administrations need to be strengthened and their rights and obligations clearly stated by law. Policy co-ordination and harmonisation by the central government help avoid inconsistencies and overlaps in the tax system. Sub-national authorities need to tap into both national and local tax revenues to provide public goods and services at local level, such as rural infrastructure to improve the transport of agricultural produce.

8. Risk management[9]

8.1. What policies and strategies has the government put in place to prevent and reduce weather, disease and price-related risks?

Agricultural investors face both production and price risks. Adverse weather, pests and diseases, and volatile prices negatively impact investment returns and result in sub-optimal investment levels. Governments can support the development of risk management instruments to help investors cope with these risks as underlined in the following questions, but they can also minimise exposure to such risks.

Agricultural production critically depends on climatic variables such as precipitation and temperatures. Consequently, climate change poses a major risk to the sector by entailing long term changes of rainfall and temperatures and increasing the frequency and severity of extreme weather events, such as droughts, floods, cyclones or storms. Furthermore, increased temperature associated with rainfall pattern changes increases the incidence of pests and diseases. The risks of being exposed to extreme weather events can be reduced in particular by implementing climate change mitigation measures. In agriculture, these include improving the management of crop and grazing land, changing land use from crops to green fallow or forests, reducing fuel consumption by increasing energy efficiency and promoting bioenergy, or sequestrating carbon in agricultural soils. Governments should thus support the development and adoption of such practices and technologies by both men and women. As for livestock, the incidence of diseases can be reduced by setting sanitary standards and establishing early warning, prevention and control systems. Several measures can also reduce the level of weather risk to which farmers are exposed, i.e. adapt to climate change, by promoting climate resilient technologies. In developing countries, these include introducing pest, drought or flood resistant crop varieties, investing in irrigation and drainage systems, promoting integrated flood management, or improving agricultural management practices to reduce for instance the risks of floods in certain watersheds by using agricultural land as water receptors.[10] Gender should be mainstreamed in these initiatives.

Price volatility represents a major concern for large investors connected to international agricultural markets as the period since 2006 has been one of extraordinary volatility on these markets. At local and national levels, such volatility may be caused by climate shocks, pests or other natural calamities, and exacerbated by poor access to technologies. The isolation of some local markets due to inadequate infrastructure can also result in high price volatility on these markets. The extent to which price changes in domestic markets mirror those on international markets depend on measures such as import duties, export taxes, non-tariff barriers or domestic policies such as price support. Large and unpredictable price volatility creates uncertainty, thereby hindering investment. Investors along the whole agricultural value chains risk losing their productive investments if price falls occur while they are locked into strategies dependent on higher price levels.

In June 2011, under the food security pillar of the G20 work on development, international organisations developed options for G20 consideration on how to reduce food price volatility. While seasonal price fluctuations and time lags in trade on domestic markets can be smoothed out by stockholding, year-to-year variations in domestic production can be more effectively and less expensively buffered by adjustments in the quantities imported or exported. Trade is an excellent buffer for localised fluctuations originating in the domestic market. Price volatility on domestic markets can also be reduced by increasing capacity to undertake frequent and systematic monitoring of the state of crops and to develop mechanisms for short-run production forecasts, using in particular satellite data and geo-information systems. Better market information and analysis can reduce uncertainties and assist agricultural investors to make better decisions.

8.2. Do private institutions or non-governmental organisations provide insurance in the agricultural sector? Does the government provide support for some types of insurance? How competitive is the insurance market in the sector?

Insurance markets are often under-developed due to information asymmetries. Asymmetric information between insurance providers and farmers can be a major source of high transaction costs and market failure. Efforts to improve insurance market functioning should thus focus on information databases. Also, information sharing arrangements can increase efficiency and open up possibilities for public-private partnerships. In particular, public insurance systems could enhance the development of databases and facilitate information sharing and access to data related to risk, coverage and indemnities to increase competition among private insurers.

In addition to developing adequate data, governments can provide insurance subsidies to address market failures. Such subsidies should not target marketable risks that can be covered by unsubsidised market-based tools. Rather, they should cover the initial costs of insurance lines with strong demand and high information asymmetries but should be gradually lowered once information asymmetries are reduced. Insurance subsidies can also help respond to catastrophic disasters, such as epizootic, for which insurance is not offered. They should then rely on explicit triggering criteria, a definition of the types and levels of assistance, and a clear delineation of the responsibility between government and producers. Such instruments may be particularly valuable in a context where climate change can lead to more frequent catastrophic disasters.

Governments can contribute to mitigating normal farming risks through tax and social security systems. They can establish income tax averaging schemes for farmers to increase the taxes paid in good years and spread losses across several years. Income deposit schemes with tax incentives to encourage income-saving in favourable years, which can be later used in lean years, can also help mitigate farm-level risks. As regards social security, general provisions can be adjusted for farmers, for instance by relaxing some of the conditions for access to income support, particularly in terms of assets.

If yield risk is systemic, developing index insurance covering risky events can reduce transaction costs. Insurance payments then depend on a publicly observable index, such as rainfall recorded on a local rain gauge. Governments may invest in weather forecast technologies and in research on appropriate indices to help set up such insurance products. Early warning systems could thus be developed using available meteorological data. The World Bank and the International Finance Corporation have already pioneered weather insurance. Since then, private and public insurance companies in India and Africa have insured more than 2 million farmers against weather risks.[11] These initiatives should be scaled up.

8.3. What mechanisms allow for the effective enforcement of forward contracts? Has the government established measures to support the development of futures markets for agricultural commodity prices?

Output market institutions play an important role in farmer risk management. Production and marketing contracts between farmers and the downstream industry, as well as the vertical integration of farmers into co-operatives or value chains, are important ways of mitigating risks as they allow farmers to transfer or share financial and market risks with other actors. A strong legal system is necessary to ensure contract enforcement, thus allowing security and flexibility in transactions. Governments can provide training to improve farmers' knowledge about marketable risks and available risk management instruments, such as contracting and futures hedging.

Co-operatives and companies offering forward contracts to farmers can transfer their own risk through futures and options markets. Farmers can then benefit from futures indirectly to reduce price uncertainty and improve price discovery. Futures need strict rules, most of which are established by the exchange boards, but appropriate financial market regulation can also contribute to their development and improve the efficiency of price hedging and discovery functions.

Increasingly, forward contracts evolve from relatively simple agreements on prices, quantity, quality and timing of sales or purchases to more diverse agreements, including production and marketing provisions, some of which may mitigate farmers' risks. Similarly, futures can combine several futures products in a single contract and are expanding towards derivatives, with OTC (over the counter) contracts that are better tailored to the needs of each trader. While large farmers with exporting interests are the most likely customers for derivatives, farmers using financial intermediaries to manage their price risk may also be able to use them.

8.4. Do agricultural extension services provide advice on co-operative arrangements among agricultural producers to help implement collective risk management strategies?

Farmers are best placed to deal with normal risks. Therefore, governments should provide knowledge and facilitate training to encourage the active engagement of farmers in the development of collective risk management strategies. In particular, they can support the development of co-operatives and industry organisations as risk management instruments. Co-operatives can pool risks across time and markets and develop payout regimes to smooth out fluctuations in returns. They can collectively engage in input and output price hedging and diversify risks through product, market and geographic diversification. Industry organisations represent a unique institutional arrangement, between their members on the one hand and between the members and the government on the other, to implement collective risk management strategies. Moreover, co-operatives can directly address farmer risks by using their authority, in particular by imposing certain production practices or marketing requirements on their members; developing and managing common funds to share the risks and costs of diseases; and serving as a single agent in contracting commercial insurance for their members. They can also develop quality assurance schemes, promote certification schemes, facilitate the adoption of sustainable farming practices and provide technical assistance to develop farmers' risk management skills, thereby reducing farmer risks.

Governments can provide the initial incentives to create mutual funds by way of start-up capital and attracting private expertise. Such mutuals, including small-scale companies, can offer coverage against specific natural crop perils. Small mutuals can address information asymmetries by having direct access to their clients and good knowledge of their members, notably through those involved in the company board. They usually benefit from a strong sense of ownership and trust amongst stakeholders. However, they may suffer from limited financial resilience due to their relatively small size and the limited scope for diversifying their risk portfolio, which can sometimes require re-insurance support from the government.

8.5. How does the government encourage diversification, including diversification in production, practices, marketing and income sources, as a risk management instrument?

Diversification has always been one of the most important risk management strategies in agriculture. It does not require the transfer of risk to other agents with imperfect information. As diversification strategies are designed by those who best know the risks, the farmers themselves, they effectively mitigate the individual risks faced by farmers. Nonetheless, diversification decisions often need to trade-off the gains in terms of reduced profit variability with losses from reduced scale economies.

Governments can encourage, in particular through research and extension, various forms of diversification, from the production of different products and the use of various marketing systems to farm and non-farm investment and income diversification. The scope of production diversification may sometimes be limited if there is a high correlation between returns from several outputs. Production practices can involve considerable differences, e.g. growing dry land and irrigated crops, as well as subtle differences, such as between some early seeded and late seeded crops. Diversification strategies also concern marketing, such as selling at different times and locations or through different channels, investing in a range of assets (different land locations, financial assets, and human capital) and conducting various on-farm and off-farm economic activities. Off-farm income can be an important source of income for farm households, thereby serving a risk management strategy and improving the management of on-farm risks and profitability.

9. Responsible business conduct

9.1. What laws and regulations govern RBC in agriculture, in particular as regards labour standards, tenure rights over natural resources, the right to health and anti-corruption and integrity standards?

Responsible business conduct (RBC) entails respect for internationally recognised standards and compliance with domestic laws and regulations, such as those on labour rights. While States have to comply with their international obligations, domestic and foreign enterprises are expected to observe and respect certain international standards derived from these obligations. To protect the collective interests of their citizens, governments should work with companies and civil society organisations, including through multi-stakeholder consultations, to promote RBC.

As regards agriculture, several RBC instruments are particularly relevant. The *OECD Guidelines for Multinational Enterprises* provide for RBC principles and standards for companies consistent with applicable laws and internationally recognised standards. The *Voluntary Guidelines on responsible governance of tenure of land, fisheries and forests in the context of national food security,* endorsed by the Committee on World Food Security (CFS) in May 2012, provide a framework for responsible tenure governance that supports food security, poverty alleviation and sustainable resource use. The CFS is also launching consultations to develop principles for responsible agricultural investment. In the agricultural sector, RBC is especially important as regards: labour rights; tenure rights of existing users of land, water and other natural resources; the right to health; animal welfare; and anti-corruption. Sustainable resource use and access to technology and innovation are also critical aspects of RBC in agriculture and are covered in other sections.

Core labour standards aim to eliminate forced or compulsory labour, abolish child labour, promote non-discrimination in employment and occupation and ensure the freedom of association and the right to collective bargaining. However, labour rights standards are frequently violated in the agricultural sector. The food and beverage industry is at the second place after the extractive industry in accusations for violations of labour rights.[12] Much of the waged agricultural employment remains informal and thus, many agricultural workers are excluded from the scope of labour laws. Women are particularly vulnerable to abuses. Respecting labour standards help to create a level playing field for all investors and improve economic performance by raising skills and providing incentives for the younger generation to accumulate human capital. Multi-National Enterprises (MNEs) are more likely to invest in countries with stricter safeguards and better protection of basic human and worker rights. Indeed, poor labour standards may damage a company's reputation, deny it access to talents from a wider pool of skills and competencies, undermine the performance of its employees, and result in missed opportunities to strengthen its competitiveness through skills development. Most countries have ratified International Labour Organization (ILO) core labour standard conventions, but compliance with, and enforcement of, the standards remains uneven.

Tenure rights over land and other natural resources are reflected in RBC standards. Land may be valued only as a productive asset while it plays a multifaceted social, cultural and religious role, provides drinking and irrigation water and can act as a safety net. Transnational land negotiations or transactions, estimated at around 83 million hectares since 2000, may lead to the displacement, the loss of livelihoods, and more limited access to land for the local population, in particular indigenous and nomadic communities. Laws should recognise and protect customary rights on land and associated natural resources. They should also provide equitable tenure rights to women whose access to land tenure is often limited. The protection of customary rights could be a precondition for negotiating land leases with large agricultural investors. Governments should also set clear and transparent rules and mechanisms to identify tenure right holders; ensure that large investors consult and negotiate with them; and fairly compensate evicted local communities.

According to the International Covenant on Economic, Social, Cultural Rights, the right to health covers timely and appropriate health care, access to safe and potable water and adequate sanitation, and healthy occupational and environmental conditions. Large agricultural investments may have direct adverse impacts on the health and safety of local communities, for instance by reducing natural buffer areas that mitigate the effects of natural hazards such as flooding and landslides, or degrading natural resources, such as freshwater. In addition, agricultural investment involves some of the most hazardous activities for workers, due to the exposure to weather, extensive use of chemicals products, difficult working postures and lengthy hours, and the use of hazardous tools and machinery. Finally, companies face risks of negatively impacting upon consumer health and safety. Well-designed and well-enforced health and environmental policies and the respect of labour standards can effectively protect the right to health.

Animal welfare is a core element of RBC.[13] An animal is in a good state of welfare if it is healthy, comfortable, well-nourished, safe, able to express innate behaviour and if it is not suffering from unpleasant states such as pain, fear and distress. Good animal welfare requires disease prevention and appropriate veterinary treatment, shelter, management and nutrition, humane handling and humane slaughter or killing.[14]

Corruption can reduce the benefits of private agricultural investments or prevent them from being realised by augmenting the cost of accessing resources and increasing the potential for conflict. Government bodies overseeing the land sector are among the public entities most affected by service-level bribery. Companies may offer undue advantages to obtain access to fertile land to the detriment of local communities holding customary land rights. Corruption may also affect the allocation of government-subsidised credit. It can increase the price of agricultural inputs as agricultural input companies can sell their products to government agencies at an elevated price to provide public officials with a share of the profit. The *OECD Convention on Combating Bribery of Foreign Public Officials in International Business Transactions (OECD Anti-Bribery Convention)* is a major instrument to prevent bribery. Clear and well-enforced laws on transparency and anti-corruption can reduce governance-related risks, thereby fostering investment.

9.2. What mechanisms are in place to enforce RBC laws and regulations effectively and ensure that local communities, particularly marginalised groups such as women and pastoralists, can: access timely and accurate information on proposed large agricultural investments affecting them; negotiate with large investors, in particular on access to land and water; ensure equitable benefit-sharing arrangements, formalised in a signed contract; receive fair and timely compensation if required; and settle contract disputes with investors?

The enforcement of RBC laws and regulations is often a challenge. It requires strong and transparent regulatory and oversight institutions with clear responsibilities and effective power as well as a well-functioning judiciary system. A lack of transparent information on planned investments increases transaction costs and deprives relevant actors of the possibility to resolve minor problems before they escalate into large conflicts. Timely and accurate information on proposed investments should thus be provided to all affected stakeholders to prevent conflicts or mistrust. Proactive consultations between government, investors and right holders should allow identifying relevant right holders and negotiating fair contracts. Such consultations should be non-discriminatory and gender-sensitive and rely on the active, free, effective, meaningful and informed participation of the various stakeholders, taking into account power imbalances between different parties (VGGT, 2012).

To ensure local acceptance and social sustainability, governments should encourage companies to share project benefits by engaging in consultations with intermediaries and local communities and by allocating monetary and non-monetary benefits to relevant stakeholders. This can also help companies select acceptable project locations and draw on local knowledge to ensure an optimal use of agro-ecological potential. Companies should also maintain written documentation of promised benefits and allocate community development funds in a transparent and open manner, particularly in the case of land acquisitions.

Agreements reached through consultations and benefit-sharing arrangements should be formalised through contracts signed between various parties that clearly define the rights and duties of all parties to the agreement. Contracting parties should provide comprehensive information to ensure that all relevant persons are engaged and informed in the negotiations, and should seek that the agreements are documented and understood by all who are affected. States and affected parties should contribute to the effective monitoring of the implementation and impacts of agreements (VGGT, 2012).

When deciding to expropriate private land and sell it or lease it to large investors, the government is responsible for providing adequate compensation to former land rights holders, including customary rights holders. To ensure fair compensation and protect their rights, local communities should be able to access grievance mechanisms at the local level as simplified and mutually beneficial ways to settle disputes with companies or governments. Properly designed and implemented grievance mechanisms increase the likelihood of resolving minor disputes quickly, inexpensively and fairly. They can help identify and resolve issues before they are elevated to formal dispute resolution methods, including courts.

9.3. Through which channels does the government communicate expected RBC standards to agricultural investors? How does it ensure a clear distinction between its own responsibilities and those ascribed to businesses?

Law-making is the key channel for communicating RBC requirements to companies. Discussions with investors and within the workplace, consultations with local communities, negotiations with trade unions, dialogue with civil society organisations and media reporting, can also play a major role in communicating such requirements. Distinction between the respective roles and responsibilities of government and business should also be clearly communicated to reduce uncertainty over RBC expectations, thus encouraging investment and allowing private and public sector actors to play mutually supportive roles in enhancing economic, social and environmental well-being. Governments need to give special consideration to SMEs as they may not have the same capacities as larger enterprises to adhere to certain policies.

9.4. How does the government support investors' efforts to comply with RBC laws and regulations? Does it support mutually beneficial partnerships between smallholders and large investors? Does it actively encourage private voluntary initiatives promoting social and environmental sustainability in the agricultural sector?

Governments can facilitate and motivate companies' efforts to comply with RBC standards by seeking out their views on laws and enforcement practices. They can provide conciliation and ombudsman facilities so that investors have the means to complain about government decisions.

They can also favour partnerships between investors and local land rights holders that do not involve large scale land acquisitions, in particular by providing practical guidance to investors or targeting investors with experience in such partnerships. Inclusive partnerships as alternatives to land acquisitions minimise investors' risks related to human rights standards. Contract farming, outgrower schemes and joint ventures can offer investors as much security of supply as direct production, spread the risks between companies and smallholders and reduce transaction costs. Contract farming can provide mutual benefits by providing farmers with better access to technology while guaranteeing a stable supply to investors. No single model emerges as the best possible option and investments may involve a combination of various models. Contract design should avoid inequitable benefit sharing. Abuses have been documented, for instance, whereby the buyer provided loans to the supplier that the latter was unable to repay, or sold inputs above market prices because of his dominant position. HRD policies should thus aim to strengthen farmers' associations and cooperatives to ensure they can negotiate with large investors on an equal basis.

Governments can also acknowledge and support private initiatives, such as the Roundtable on Sustainable Palm Oil (RSPO) formed in 2004 to promote palm oil development through credible global standards and engagement of stakeholders.

9.5. Does the government participate in inter-governmental consultations to promote recognised RBC concepts and principles?

Governments should co-operate with each other and with other actors to strengthen the international legal and policy framework in which business is conducted. They can draw from internationally recognised RBC principles and standards to develop their own legislation. Increased global acceptance of common RBC principles also helps to ensure that home countries better promote RBC, thus facilitating RBC enforcement in host countries.

Multilateral instruments dealing with RBC in agriculture include in particular:

- the OECD Guidelines for Multinational Enterprises;
- the Voluntary Guidelines on responsible governance of tenure of land, fisheries and forests in the context of national food security;
- the International Labour Organization (ILO) Tripartite Declaration of Principles concerning Multinational Enterprises and Social Policy;
- the United Nations Global Compact; International Finance Corporation Performance Standards on Environmental and Social Sustainability;
- the Akwe Kon Voluntary Guidelines for the conduct of cultural, environmental and social impact assessments regarding developments proposed to take place on, or which are likely to impact on, sacred sites and on lands and waters traditionally occupied or used by indigenous and local communities;
- Large-scale land acquisitions and leases - A set of core principles and measures to address the human rights challenge of the UN Special Rapporteur on the Right to Food; and
- the Guidelines on Business, Land Acquisitions and Land Use of the Institute for Human Rights and Business.

10. Sustainable use of natural resources and environmental management

10.1. Do existing environmental policies, laws and regulations effectively ensure a sustainable use of natural resources, in particular by setting clear environmental standards, requiring independent environmental impact assessments and ensuring that the pricing of natural assets reflect their true scarcity value? Do they take into account the specificities of the agricultural sector?

Poor agricultural practices, such as the cultivation of unsuitable land and the inappropriate or excessive use of agricultural technologies and chemicals, can lead to the over-exploitation and degradation of natural resources, particularly land degradation, water resource depletion and pollution and deforestation, thereby reducing agricultural productivity and threatening long-term food security. Natural resource depletion has a cost to society and the economy as a whole and can deprive entire communities from their livelihoods. Water resource depletion and pollution is of a particular concern in light of the need for increased agricultural production. At present, agriculture is the largest user of water world-wide and accounts for about 70% of the total global freshwater withdrawals. An increasing number of countries or parts of countries are reaching critical levels of water scarcity. Agriculture is also a point source of water pollution, for example, from intensive livestock farms and the disposal of residual pesticides. Water pollutants from agriculture include nitrogen runoff and leaching into water systems from using and disposing of nutrients and pesticides and soil sediments.[15] It should also be noted that reducing food losses and waste is a key aspect of a more sustainable food and agriculture system. A recent FAO study suggests that roughly one-third of food produced for human consumption is lost or wasted globally, meaning that significant amounts of the resources used in food production are also wasted.[16]

Policy-makers face the challenge of creating the right incentives to optimise resource use from an economic, environmental and social perspective. Sustainable resource management would allow agricultural investors to maximise returns on their investment by harnessing long-term economic benefits. Environmental policy should promote sustainable farming practices, such as agro-forestry and agroecology, as well as resource use efficiency to increase production relative to inputs used, and ensure that prices reflect the scarcity value of natural resources and the cost of environmental impacts.

Environmental policy should also aim to: reduce environmentally harmful subsidies; enshrine the polluter pays principle within the legal and regulatory framework; oblige investors to internalise the costs of environmental degradation by making environmental impact assessments compulsory, issuing environmental permits and sanctioning environmentally damaging activities; provide incentives for the supply of environmental goods and services and encourage participatory management of natural resources; and reduce waste and post-harvest losses. As free access to resources can encourage over-exploitation, secure tenure rights can help ensure optimal resource use, in particular for land and forests. Developing a water management policy and undertaking regular reviews of the allocation of water rights can help ensure an equitable access to and sustainable use of water resources. Governments can also provide information and trainings and promote strong certification schemes and marketing transparency to consumers to influence the environmental performance of agriculture.

Mainstreaming ratified international treaties and agreements on environmental management into national policies and laws can help promote such measures and ensure their effective implementation. Furthermore, international initiatives, such as the carbon market mechanism REDD (Reducing Emissions from Deforestation and Forest Degradation) launched by the UN in 2008, and private voluntary initiatives can be supported and implemented.

10.2. What institutional mechanisms allow for the effective implementation, monitoring and enforcement of environmental policies, laws and regulations?

Effective institutional mechanisms, in particular strong regulatory and oversight institutions with well-defined responsibilities and effective power, should be in place to ensure that environmental management principles are properly implemented and sanctions are well enforced. The efficient, sustainable and equitable management of land and water resources requires an integrated approach involving multiple stakeholders including Ministries, public agencies, sub-national authorities and private actors. This remains a challenge due to the high degree of fragmentation of roles and responsibilities in many countries.

The issuance of environmental permits can help promote sustainable investment. Such permits should be issued in a timely and transparent manner. Environmental impact assessments should be conducted by independent parties and sanctions for non-compliance with environmental legislation effectively applied by a well-functioning judiciary system. All stakeholders affected by negative environmental impacts, in particular marginalised groups, should be able to raise complaints and receive timely, adequate and effective compensation. The government can facilitate or mediate disputes related to investment in agriculture, provide administrative support and help negotiate compensation.

10.3. Do existing policies promote access to clean, energy-efficient and low input technologies and encourage their adoption by large and small agricultural investors? Do they encourage investment in technologies using agricultural waste as an energy feedstock?

Access to clean and energy-efficient technologies is particularly important in agriculture as resource scarcity is increasing and as the sector is an important contributor to greenhouse gas emissions. Indeed, the agricultural sector is the 4th largest sectoral contributor to such emissions after energy, industry and forestry including deforestation. The intensification of agricultural practices, especially the use of fertilisers in developing countries, and the change of dietary patterns towards increased meat consumption are projected to lead to increased methane and nitrous oxide emissions by 2050.[15] Access to clean and energy-efficient technologies is even more important in a context of climate change. While agriculture can contribute to climate change, agricultural production is particularly vulnerable to climate change consequences, such as global warming, rising sea levels, changing precipitation patterns and extreme weather events.

Energy policy should take into account the sector's needs. Bringing down trade barriers in environmental goods and services, attracting investors experienced in using clean and energy-efficient technologies and encouraging R&D on these technologies would facilitate their dissemination and use, thereby increasing local agriculture competitiveness and improving the environmental sustainability of productive activities. Such efforts need to be accompanied by measures to increase the capacity of SMEs to absorb these technologies (see human resources section).

The production of biofuels has been promoted as a source of clean energy in several countries over the last decade. The economic and environmental soundness of such production should be carefully assessed as its economic viability remains uncertain in many countries. The inflexible nature of mandates may result in market distortions. Moreover, the impact of biofuels on the reduction of greenhouse gas emissions, especially when produced with palm oil or jatropha, can be negative if previously non-cultivated areas are cleared for biofuel production.

Notes and references

1. The issue of umbrella clauses is discussed in OECD (2009), "Interpretation of the Umbrella Clause in Investment Agreements" in *International Investment Law, Understanding Concepts and Tracking Innovations.*

2. www.apia.com.tn

3. Chai and Goyal (2008), Tax Concessions and Foreign Direct Investment in the Eastern Caribbean Currency Union, IMF Working Paper, November 2008.

4. Brooks, J. (ed.) (2012), *Agricultural Policies for Poverty Reduction,* OECD Publishing. doi: 10.1787/9789264112902-en

5. Idem.

6. OECD (2013), *Policy Guidance for Investment in Clean Energy Infrastructure: Expanding Access to Clean Energy for Growth and Development*, Report submitted to the G20 with inputs from the World Bank and the United Nations Development Programme.

7. Ceylon tea and its Lion logo were developed by the Sri Lanka tea board.

8. OECD (2013), *Bribery and Corruption Awareness Handbook for Tax Examiners and Tax Auditors*, OECD Publishing. doi: 10.1787/9789264205376-en

9. For further information, see OECD (2011), *Managing Risk in Agriculture: Policy Assessment and Design*, OECD Publishing. doi: 10.1787/9789264116146-en

10. OECD (2014), *Climate Change, Water and Agriculture: Towards Resilient Systems*, OECD Studies on Water, OECD Publishing. doi: 10.1787/9789264209138-en

11. World Bank (2009), *Innovative Finance for Development Solutions,* Washington DC.

12. European Commission (2011), Report - A sectoral approach to CSR to tackle societal issues in the food supply chain, High level forum for a better functioning food supply chain, expert platform on the competitiveness of the agro-food industry.

13. Animal welfare means how an animal is coping with the conditions in which it lives. It refers to the state of the animal. The treatment it receives is covered by other terms such as animal care, animal husbandry and humane treatment.

14. OIE (2013), Terrestrial Animal Health Code, World Organisation for Animal Health.

15. OECD (2010), *Sustainable Management of Water Resources in Agriculture*, OECD Studies on Water, OECD Publishing. doi: 10.1787/9789264083578-en

16. OECD/FAO (2012), *OECD-FAO Agricultural Outlook 2012*, OECD Publishing. doi: 10.1787/agr_outlook-2012-en

ORGANISATION FOR ECONOMIC CO-OPERATION AND DEVELOPMENT

The OECD is a unique forum where governments work together to address the economic, social and environmental challenges of globalisation. The OECD is also at the forefront of efforts to understand and to help governments respond to new developments and concerns, such as corporate governance, the information economy and the challenges of an ageing population. The Organisation provides a setting where governments can compare policy experiences, seek answers to common problems, identify good practice and work to co-ordinate domestic and international policies.

The OECD member countries are: Australia, Austria, Belgium, Canada, Chile, the Czech Republic, Denmark, Estonia, Finland, France, Germany, Greece, Hungary, Iceland, Ireland, Israel, Italy, Japan, Korea, Luxembourg, Mexico, the Netherlands, New Zealand, Norway, Poland, Portugal, the Slovak Republic, Slovenia, Spain, Sweden, Switzerland, Turkey, the United Kingdom and the United States. The European Union takes part in the work of the OECD.

OECD Publishing disseminates widely the results of the Organisation's statistics gathering and research on economic, social and environmental issues, as well as the conventions, guidelines and standards agreed by its members.

OECD PUBLISHING, 2, rue André-Pascal, 75775 PARIS CEDEX 16
(20 2014 07 1 P) ISBN 978-92-64-21269-5 – 2014-01